The "King Tiger" Vol.II

• Development • Units • Operations •

Wolfgang Schneider

Cover:
A Tiger II during a hard fight northeast of Aachen, late November 1944.

Inside Cover:
Of the six existing King Tigers still in Europe (Bovington, la Gleize, Munster, Saumur, Shrivenham, Thun), the one in the French Tank Museum (Saumur) is the only one still in driveable condition. The paint is not completely accurate, the cross is too wide, the number 233 (Heavy Panzer Battalion 506) is some 20% too big.

Schiffer Military History
Atglen, PA

Photos were provided by:
Bovington Tank Museum
Egon Kleine
La Gleize Museum
Dr. Franz Lochmann
Panzermuseum Munster
Jobst von Römer
Richard Freiherr bon Rosen
Erwin Stahl
Thyssen Henschel, Kassel
Wilhelm Winkelmann
The other photos came from the author's archives

Sketches were provided by:
The magazines *PANZER* and *Tank Magazine* of Tokyo.

Translated from the German by Dr. Edward Force
Central Connecticut State University.

Library of Congress Catalog Number: 89-84175

ISBN: 0-88740-287-9
Printed in China

This book originally published under the title, *Der Königstiger 2.Band,* by Podzun-Pallas Verlag, GmbH 6360 Friedberg (Dorheim), © 1988. ISBN: 3-7909-0336-1

Published by Schiffer Publishing Ltd.
4880 Lower Valley Road
Atglen, PA 19310
Phone: (610) 593-1777; Fax: (610) 593-2002
E-mail: Info@schifferbooks.com
Please visit our web site catalog at

In Europe, Schiffer books are distributed by
Bushwood Books
6 Marksbury Avenue Kew Gardens
Surrey TW9 4JF England
Phone: 44 (0) 20-8392-8585; Fax: 44 (0) 20-8392-9876

This book may be purchased from the publisher.
Include $3.95 for shipping. Please try your bookstore first.
We are always looking for people to write books on new and related subjects.
If you have an idea for a book please contact us at the above address.
You may write for a free catalog.

FOREWORD

With the help of former members of troop units equipped with King Tiger tanks it has been possible to gather a series of particularly interesting photos. This volume was ultimately assembled from approximately 400 Tiger II photos. Almost exclusively, photos rarely published before or published for the first time here were used. For the first time, a complete listing of all King Tiger tanks delivered to the troops is provided (page 27). The presentation is completed by cutaway line drawings and a list of the heavy tank units. The editors and the author would be very grateful for any further contacts with former "Tiger" men (for example, sPz.Abt. (Fkl. 301) that might result from this book.

Technical Evaluation

Ever since the war, the controversial question has been debated as to which Wehrmacht tank was the most refined, the most powerful of all. In this context the Panther and the Tiger II are always named. As so often, no absolute decision clearly favoring the one or the other type can be made here.

The Panther presented a particularly successful combination of all three fighting-value parameters: firepower, mobility and armor protection. For the conditions of those times it was very mobile, well-armored (designed to deflect shots), and armed with a high-performance 7.5-cm tank gun.

As for the Tiger II, the superior firepower of the 8.8-cm L/71 tank gun and the heavier armor protection can be cited. Naturally, these were paid for by a great increase in overall weight with the same engine power. This high weight caused serious problems in combat, especially for the components of the running gear, which caused many breakdowns and greater stress. As with the Panther, the side panels of the Tiger II were weak points and were redesigned in the process of manufacturing, and the existing tanks were reequipped accordingly. In addition, the swing arms tended to bend. These and the lack of firmness of the rubber-mounted road wheels caused their angled positions or axial displacement. This in turn influenced the course of the track links. The links themselves were under a considerably higher internal pressure, which caused the link bolts to bend.

The first fifty King Tigers were equipped with the so-called Porsche turret (above). This was more sharply sloped and had a very rounded front. Its manufacture was very expensive, and it also turned out that the shape essentially trapped shots and caused a particularly great danger to the turret turning circle. Thus Krupp produced a turret that offered greater armor strength and eliminated these problems, as well as being much simpler to produce. With this so-called production turret all subsequent Tiger II tanks were equipped. The two photos offer a comparison of the two types.

During series production, a number of improvements and armament changes were introduced. One example is the exhaust system. Originally the exhaust pipes were jacketed (above) for sound damping and better heat conduction; later vehicles had open tailpipes.

A great number of technical breakdowns resulted. All in all, the performance weight of the Tiger II was too low. The tactical mobility was also diminished decisively because of the high fuel consumption caused by the heavy weight of the vehicle.

Thus the great strength of the King Tiger was its armor plate, which made it "immune" to the greater part of the shells that hit it. In fact, only a small number of the Tiger II tanks fell victim to direct enemy action. The overwhelming majority of them had to be abandoned and/or blown up by their crews — because they were immobile. The Tiger II brought its strength directly to bear in tactical situations in which superior numbers of enemy tanks had to be driven off, or when a decision had to be forced quickly at a focal point. The Panther was better suited to large-scale or fast-moving operations.

In terms of its design concept, the firepower of the Tiger could have been utilized better if independent units had not been formed (which were then divided all too often); a better solution would have been combining the heavy tanks in units with the more mobile light tanks (for example, a heavy company in a tank regiment).

Because many more of them existed, the Panther was of particular value to the Panzer troops; the Tiger excelled particularly in tank-against-tank duels and in the numerous crisis situations of the last year of the war.

A series of tanks served for testing. This photo shows V8 in a rescue maneuver at Kummersdorf. Notice the number plate — not used in combat — on the bow plate.

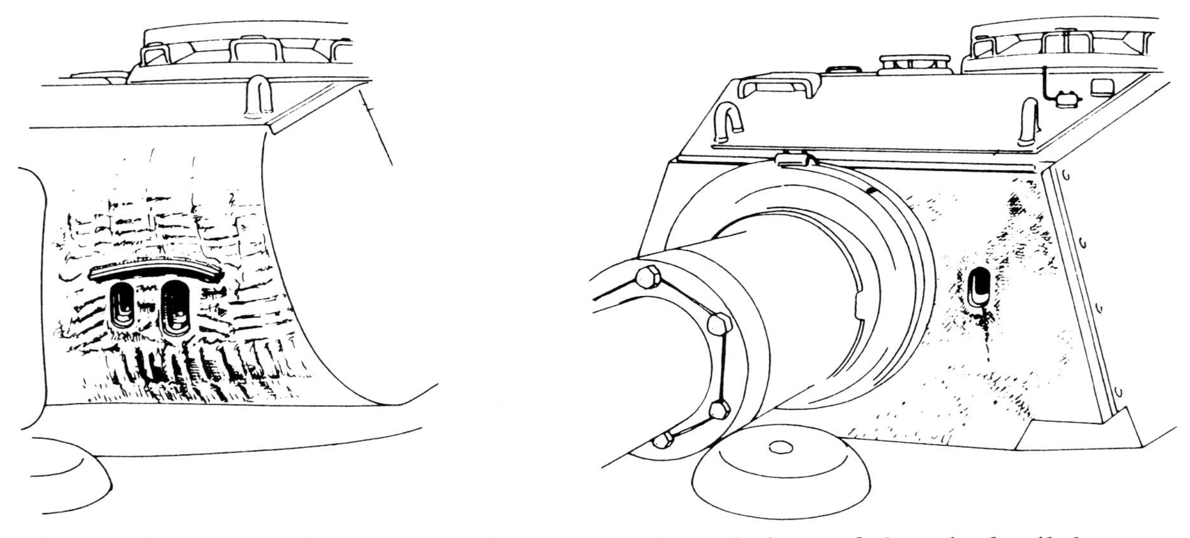

The two drawings make clear the different turret front design and show in detail the ports of the turret aiming telescope to the left of the gun mount. The wire pin located between the ring and the commander's cupola on the production turret was a makeshift aiming aid for the commander during observation from the opened hatch; with it he could take rough aim at an enemy target, and the gunner could recognize it in his optic's field of vision.

Thus the tactical significance of both tanks was very great, but the legendary reputation of the King Tiger was originated mainly by the tank troops of the Allies or the crews of the antitank weapons, who had to recognize their own inferiority or take a chance against this Colossus from short range.

One should also not neglect to consider the fact that the manufacture of the Tiger was considerably more costly in terms of both labor and materials, and that the aim was to go beyond the Panther II to establish a single main battle tank. As is well known, the end of the war put a stop to the prevailing "tank dualism." It should also be mentioned that the carrying abilities of most roads and cross-country areas, and in particular the greater part of the bridges at that time, were hopelessly overburdened (and by the Panther too).

Thus the technological role of the Tiger II can be evaluated more highly in the end than its actual value in combat.

Left:
This photo shows the first production tank (note number 001 on the left towing eye), It is clear to see that the commander's hatch required a bulge on the left side of the turret wall.

Right:
Inside view of the front of the hull, with the driver's seat and the radioman's seat to the right beside it. In between is the gearbox. Note the quarter-circle steering wheel.

Lower right:
This picture shows the port for the short-range defensive weapon from outside, in both closed and opened (right) conditions.

Below:
In the turret roof ahead of the loader was the short-range defensive weapon, which was operated from inside. Fog cartridges were generally utilized; explosive shells were also developed but were used very seldom.

CUTAWAY DRAWING OF THE KING TIGER

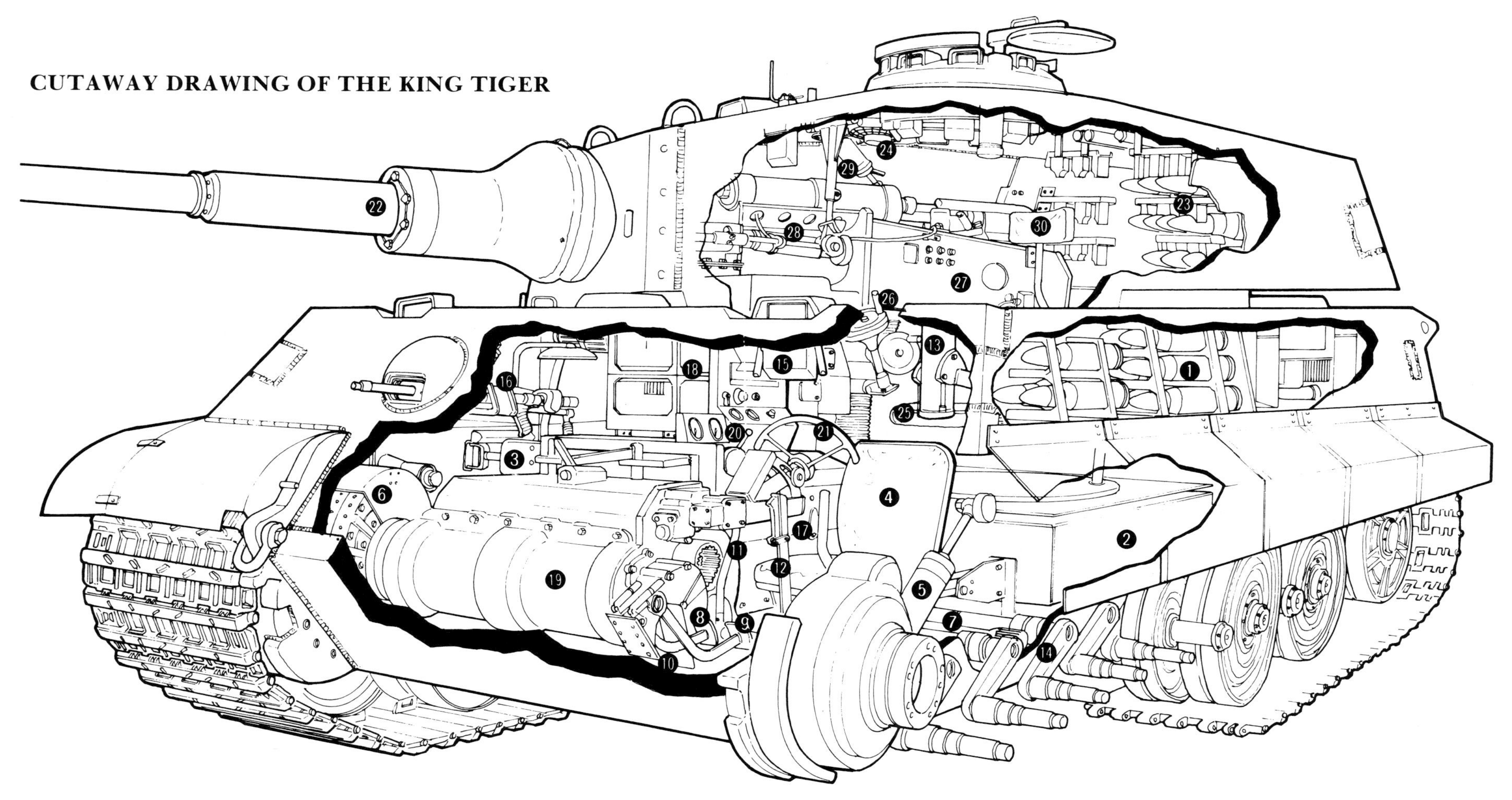

BATTLE TANK TIGER II
(Special Motor Vehicle 182)

DETAILS SHOWN

1. Left ammunition rack
2. Turntable
3. Radioman's seat
4. Driver's seat
5. Track tension cylinder
6. Disc brakes
7. Torsion bars
8. Right brake pedal
9. Left brake pedal
10. Drive pedal
11. Lateral gearing
12. Hand brake lever
13. Elevation crank
14. Swing arms
15. Driver's periscope
16. Bow machine gun
17. Gearshift drive
18. Radio set
19. Gearbox
20. Gearshift lever
21. Steering wheel
22. Tank gun
23. Rear turret ammunition rack
24. Ventilator
25. Gunner's seat
26. Traverse aiming wheel
27. Gun mount with baffle
28. Turret aiming telescope
29. Close-combat weapon
30. Commander's seat

CUTAWAY DRAWING OF THE TURRET

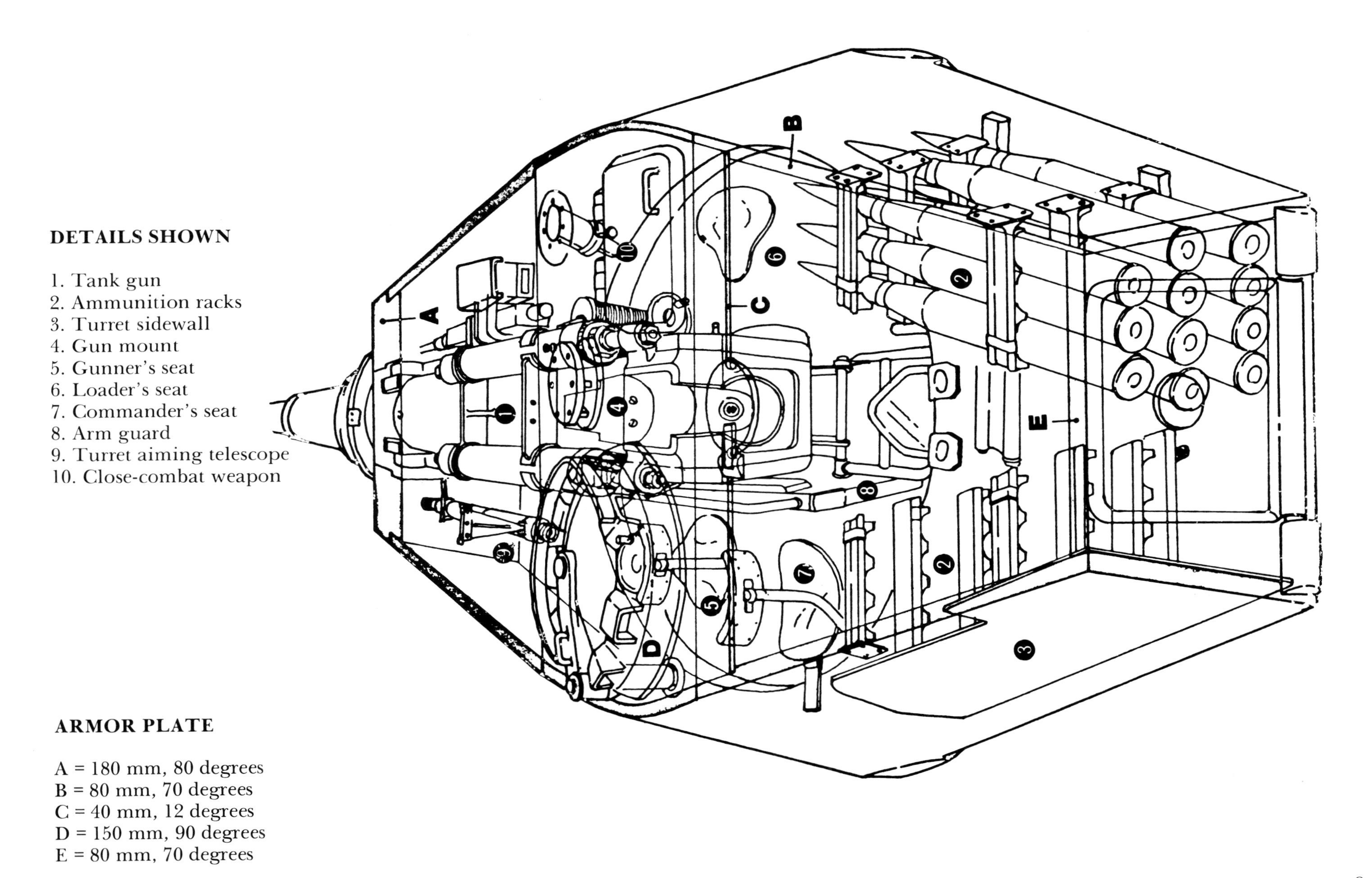

DETAILS SHOWN

1. Tank gun
2. Ammunition racks
3. Turret sidewall
4. Gun mount
5. Gunner's seat
6. Loader's seat
7. Commander's seat
8. Arm guard
9. Turret aiming telescope
10. Close-combat weapon

ARMOR PLATE

A = 180 mm, 80 degrees
B = 80 mm, 70 degrees
C = 40 mm, 12 degrees
D = 150 mm, 90 degrees
E = 80 mm, 70 degrees

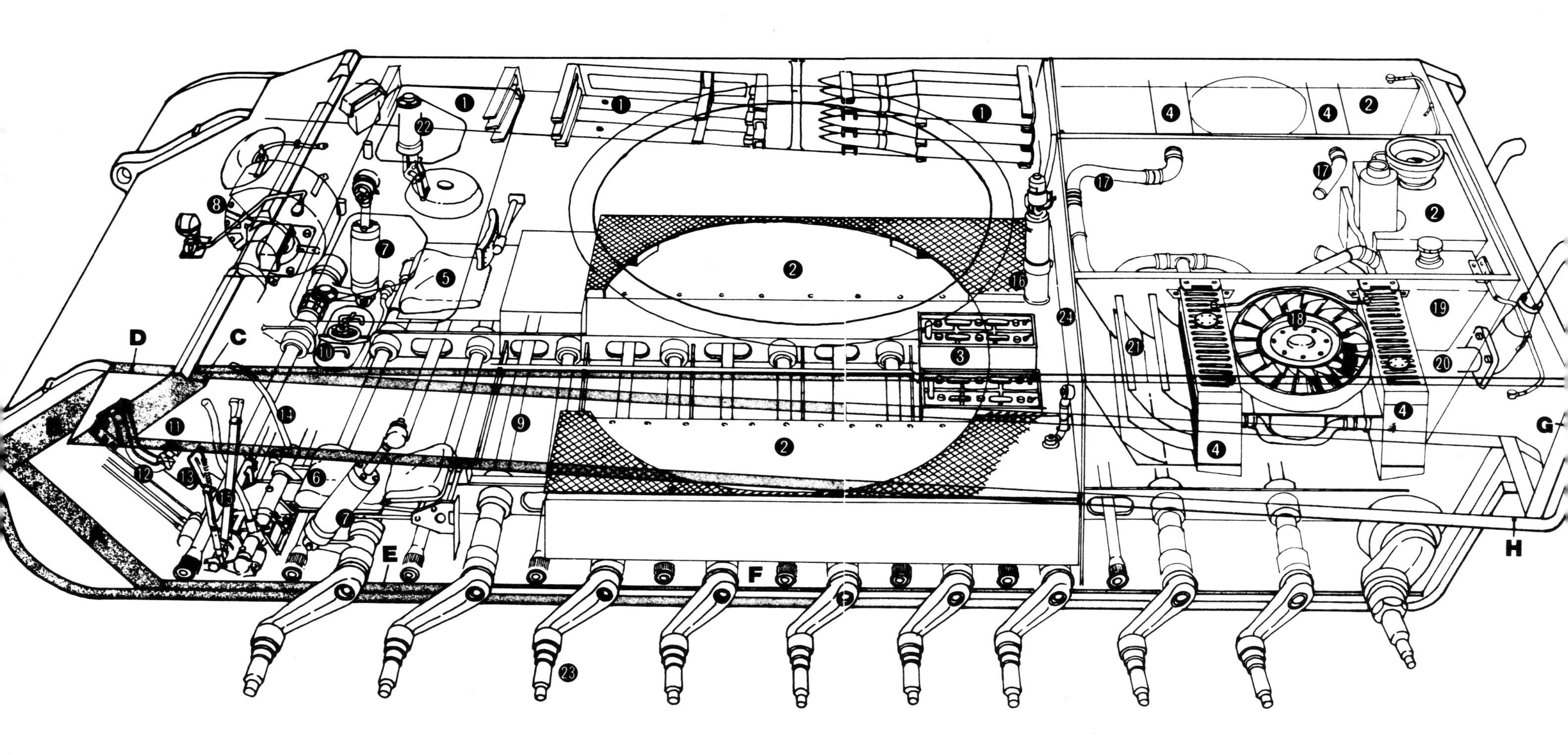

CUTAWAY DRAWING OF THE RUNNING GEAR AND HULL

DETAILS SHOWN

1. Hull ammunition racks
2. Turntable
3. Batteries
4. Radiator
5. Radioman's seat
6. Driver's seat
7. Track tension cylinder
8. Disc brakes
9. Torsion bars
10. Emergency exit hatch
11. Right brake pedal
12. Left brake pedal
13. Drive pedal
14. Emergency control lever
15. Hand brake lever
16. Fire extinguisher
17. Radiator duct
18. Ventilator wheel
19. Coolant storage tank
20. Exhaust pipe
21. Radiator baffles
22. Radioman's hatch
23. Swing arms
24. Engine room bulkhead

ARMOR PLATE

A = 100 mm, 40 degrees
B = 150 mm, 40 degrees
C = 40 mm, 0 degrees
D = 80 mm, 65 degrees
E = 40 mm, 0 degrees
F = 25 mm, 0 degrees
G = 80 mm, 60 degrees
H = 80 mm, 65 degrees

Heavy Panzer Battalion 503/ Feldherrnhalle

The First Company of this unit was the first troop unit to be supplied, on June 6-11, 1944, with the Tiger II and transferred to France. They were unloaded 80 kilometers west of Paris (in Dreux), so that about 200 kilometers had to be covered (in four nights) to the area east of Caen, which put extraordinary strains on the vehicles and induced a number of technical breakdowns. Only on July 18 was an attack made in the direction of Cagny, and enemy attempts to break through were turned back.

At the beginning of August, the 3rd Company was also rearmed with the Tiger II at Mailly-le-Camp and went into action north of the Seine; in the Amiens area the last tank had to be blown up in the tangle of retreat fighting.

Immediately after the Allied invasion on the Normandy coast of France, the 1st Company of Heavy Panzer Battalion 503 was supplied with the King Tiger. The two photos show one of the tanks from the early production lot with the simple aiming telescope port (later two) at left beside the tank gun, which is still extended in one piece. At the left side of the turret the welded opening of the original ammunition hatch can be seen.

In the woods near Canteloup the tanks of the unit prepare for their combat; in the foreground is the tank with turret number 114.

Right:
During the bitter fighting east of Caen, all the tanks were lost one by one. The picture shows one on the road to Vimoutiers.

Below and lower right:
Both photos show a disabled Tiger II in Plessis-Grimoult. In a later attempt to tow it away, the turret fell off to the side (right).

Left:
Another tank was abandoned while still on a towline from a Panther recovery tank, and is being passed by Canadian troops (in the foreground is a "Staghound" armored scout car).

Lower left:
General Eisenhower examines a capsized "Royal Tiger." The floor of the fighting compartment was bent by an explosion. The arrangement of the servicing openings and the emergency exit hatch at the left front can be seen.

Below:
This drawing shows all the openings in the base of the Tiger II.

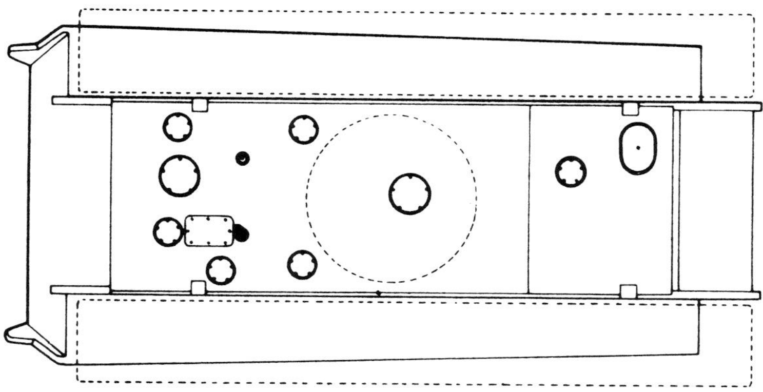

The 3rd Company turned its remaining tanks over to the 2nd and was likewise supplied with Tiger II tanks in Mailly-le-Camp. This photo shows part of the company there during firing drill.

On August 11, 1944 the first five railroad cars were loaded and rolled down the tracks to the front. The next day the train was attacked by five Thunderbolts near Esterney. The platoon leader's tank of Leutnant Freiherr von Rosen (311) caught fire, fell onto the tracks during emergency unloading and tipped over.

Below: The company was unloaded in Paris, drove along the Champs Elysees and stopped briefly at a rest area in the Bois de Vincennes.

The unit was completely reorganized that September in Paderborn-Sennelager, and on October 12 it was loaded on a train for transport to Hungary. There they first undertook the disarming of the renegade Hungarian troops.

Next the unit was subordinated to the 24th Panzer Division for an attack east of Szolnok. An important partial success was the prevention of an advance into the back of the Wöhler Army Group. In the time that followed, the unit saw action repeatedly in the area east of Budapest, then south of it.

In December the unit was transferred to the Stuhlweissenburg area and played an important part in stabilizing the threatening situation there. On December 21 it was renamed "Feldherrnhalle." The bitter fighting in the Stuhlweissenburg area lasted until early February 1945. Almost all the unit's tanks were lost, so that six replacement tanks had to be delivered in February. Fighting began around the Gran bridgehead, the removal of which delayed the Russian preparations to attack in the direction of Pressburg and Vienna. Despite another delivery of tanks, though, the unit only had a strength of 21 tanks (21 companies).

Heavy fighting on the east side of the Little Carpathians and in the Zistersdorf oil fields pushed the unit back in the direction of Budweis. The last two Tiger II tanks were blown up on May 10. The majority of the unit fought its way through to the Americans but was turned over to the Russians.

After its reconstitution, the unit was sent to Hungary and first used to put down renegade Hungarian troops in Budapest. The 3rd Company (below) took up a position below the castle.

This photo shows a tank of the 2nd Company (231) in the upper city. A Hungarian "Nimrod" anti-aircraft tank is in position.

The number one tank of the 2nd Company (200) rolls over one of the makeshift barricades.

This unique series of photos shows the development of a battle for the 1st Company on December 23, 1944 at Polgardi (east of Stuhlweissenburg, Hungary). At first the tanks pick up ammunition for their guns (left), then the attack moves through the village (lower left) with one tank on each side of the street (here on the right). Below: The King Tiger in the left foreground burns after taking a direct hit.

Right:
An antitank gun on the left flank that had not been silenced hit Tank 124 at the shortest range. The following tanks guard the crew as they abandon the tank.

Below:
After the village has been taken, the extent of the loss becomes clear. The close-up photo shows the burning tank which suffered a hull penetration at the left front. The driver and radioman lost their lives.

Left: Amazing but true: the unit was reissued the same two Tiger II tanks (with Porsche turrets) that they had left behind at Mailly-le-Camp.

Above and below: The other photos show additional tanks of the unit in the Stuhlweissenburg area. Above is the chief tank of the 1st Company (100), at the lower left another 1st Company tank, below No. 314.

Left:
Tank 333 tows a damaged comrade; 311 goes ahead.

Lower left:
Numerous damaged tanks were taken to Vienna by train and sent back in January and February after being repaired.

Below: A King Tiger and its crew at the burial of a comrade.

The Heavy Panzer Battalion 501 unloading its tanks at Kielce, Poland. Tank 313 is being pushed by a second Tiger onto the right-side combat track laid out ahead of it. The narrow shipping track is still on the left side. Then the track aprons will be screwed on.

Heavy Panzer Battalion 501

After the beating they took in the defeat of the Army Group Center, this unit was reconstituted at the Ohrdruf (Thuringia) training camp in July of 1944 and sent on the march to Poland at the beginning of August. From the unloading depot they marched some fifty miles through the country to the front. On the way a large percentage of their tanks suffered technical problems. The rest took part in an attack across the Czarne with the 16th Panzer Division. The goal was to put pressure on the bridgeheads at Baranov and Sandomierz — but they did not succeed. Fighting in the Ostrowice area, and later by Radom, followed.

On December 21 the unit was renumbered as Heavy Panzer Battalion 424 and, at the beginning of January, it was deployed further to the south, in country unfavorable to tanks, and supplied with Tiger I tanks from the recently disbanded Tank Unit 509.

The vigor of the major Soviet attack of January 12 exceeded all fears, and within a few days the unit was almost completely smashed.

The survivors of this unit, which had been shattered once again, gathered in the Sorau area and formed the nucleus of Jagdtiger Unit 512.

Above:
Under what precarious conditions the repair crews had to work is shown in this picture of a tank of the 3rd Company (314) being repaired outdoors in swampy country. This unit did not have the usual rounded upper loop of the "3" on the turret number, but in straight lines (see below). Several modelmaking firms mistakenly turned it into a "7" later.

The platoon leader's crew of the 1st Platoon of the 2nd Company (211) cleaning and servicing the tank.

DELIVERIES OF THE PANZER VI B "KING TIGER" TANK

Datum	Bedarfsträger	Zahl
Ende 1943	Versuchsbetrieb	3
1. 4. 1944	ErprStelle Kummersdorf	1
9. 5. 1944	"	3
2. 6. 1944	"	3
23. 2. 1945	"	3
1. 4. 1944	PzErsatz- u. AusbAbt 500	2
30. 6. 1944	"	4
9. 7. 1944	"	1
10. 8. 1944	"	2
6. 1. 1945	"	1
14. 4. 1944	PzKp (Fkl) 316	5
25. 8. – 21. 10. 1944	PzAbt (Fkl) 301	31
25. 6. – 7. 8. 1944	PzAbt 501	45
12. 6. 1944	PzAbt 503 (1. Kp)	12
27./29. 6. 1944	" (3. Kp und Stab)	14
19. – 22. 9. 1944	"	45
11. 3. 1945	PzAbt FHH	5
26. 7. – 29. 8. 1944	PzAbt 505	45
20. 8. – 12. 9. 1944	PzAbt 506	45
8. 12. 1944	"	6
13. 12. 1944	"	6
(30. 3. 1945	" von SS-PzAbt 501	13)
9. 3. 1945	PzAbt 507	4
22. 3. 1945	"	11
28. 9. – 3. 10. 1944	PzAbt 509	11
5. 12. 1944 – 1. 1. 1945	"	45
30. 1. 1945	PzAbt 511	3
28. 7. – 1. 8. 1944	SS-PzAbt 101	14
17. 10. 1944 – 22. 1. 1945	SS-PzAbt 501 (101)	40
27. 12. 1944 – 6. 3. 1945	SS-PzAbt 502	37
19. 10. 1944 – 16. 1. 1945	SS-PzAbt 503	33
(20./21. 10. 1944	" von PzAbt 301	10)
	Ausgeliefert	477

STRUCTURE AND STRENGTH OF A HEAVY 'TIGER' TANK UNIT "TIGER"

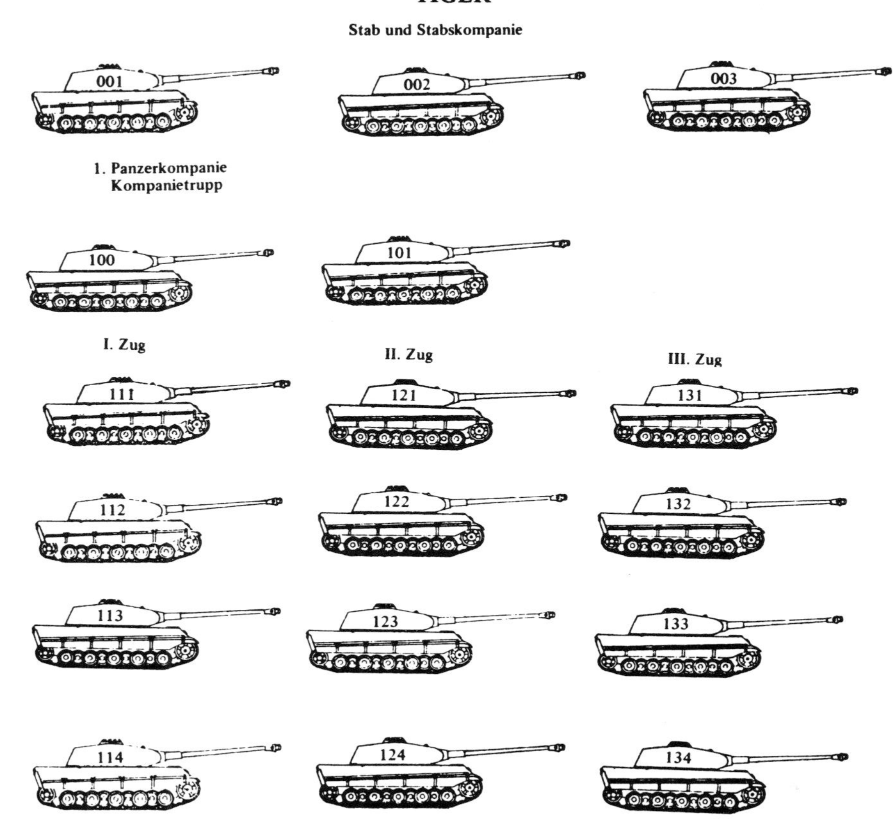

Notes:
At times there were departures from the usual turret numbering system (for example, numbering the three staff tanks I, II and III).
The 2nd and 3rd Tank Companies had the same structure as the 1st Company, with turret numbers beginning with 2.. and 3.. respectively.

Strength	Officers/NCOfficers/Men = Total
Staff and Staff Company	9/37/130 = 176
Supply Company	5/55/188 = 250
Repair Company	3/37/162 = 207
1st Tank Company	4/46/38 = 88
2nd Tank Company	4/46/38 = 88
3rd Tank Company	4/46/38 = 88
Total strength	29/267/594 = 897

Tank Unit 505 was reorganized in Ohrdruf, Thuringia, and first carried out intensive individual and group training. The numbering of the tanks differed from the usual norm. The tactical number is found on the side of the gun jacket and on the cover. Instead of this, the emblem of the unit, a knight armed with a lance, was painted in bright colors on both sides of the turret.

Heavy Panzer Battalion 505

In August and September of 1944 this unit was refreshed at Ohrdruf, equipped with Tiger II tanks, and loaded up for transport toward Nasielsk on September 8.

Subordinated to the 3rd Panzer Division. it was assigned to remove the Russian bridgehead over the Narev. The attack was broken off, though, and the unit was transferred by train to the Wirballen area to join the XXVII. Army Corps.

On October 16 the defensive fighting against Soviet attacks in the Eydtkau area began, followed by action in the Kassuben area under the "Hermann Göring" Panzer Division as of October 20.

505

Early in November the unit fought with the Führer-Begleit-Brigade, and then served as army reserve in the Goldap area. It had a rest until well into December, and repair work succeeded in bringing the unit back to about two-thirds of its strength. Then the unit fought well in the battles south and west of Königsberg. Before Pillau at the end of April the unit had to blow up its six remaining Tiger II tanks. Two tanks still saw action at Fischhausen and were blown up by their crews after they broke down.

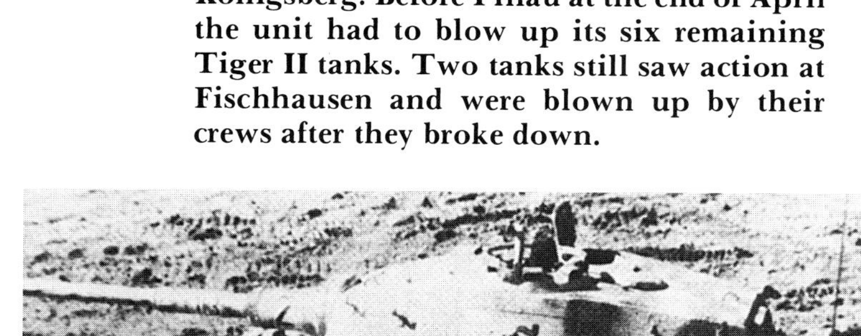

Above:
The show must go on (this came in question rarely enough)! In the background is the platoon leader's tank of the First Platoon, First Company (111).

Upper right:
After a few days in battle, the tanks no longer made such a good appearance. The dents made by shells that hit the bow of 211 make this clear.

Right:
The time spent in railroad transport was also utilized by the crews. Here a soldier of the 1st Company shows what a good washline the tank gun makes.

Above: The newly established Heavy Panzer Battalion 506 saw service only in the west. It first received five King Tigers with the old Porsche turret. This photo shows them being transported by rail to the war zone near Aachen.

Above: After their victorious combat at Arnheim, the unit fought northwest of Aachen. Here Tank 111 fell into American hands near Gereonsweiler. The photo shows it with white stars on cloths; on the gun cover is the chalked inscription "Warning, mined."

Heavy Panzer Battalion 506

This unit was also refreshed at Ohrdruf in August and September of 1944 and rearmed with Tiger II tanks. At the end of September it was transported by rail to the area west of Arnheim, where the unit fought in the breaking of the Allied "Market Garden" air landing. Next it was transferred to the Gereonsweiler area and took part in the fighting around Aachen.

Early in December the unit was refreshed in the Grevenbroich area and strengthened with the heavy tank company "Hummel", armed with Tiger I tanks.

In mid-December it was transferred into the Eifel Mountains and then saw action in the Bastogne area, where the unit thwarted numerous US relief actions.

In February of 1945 divided action took place in the Schnee-Eifel and west of Prüm.

Early in March it fought at Kyllburg and Boxberg. After that the unit's last two tanks had to be blown up.

Without tanks, the remainder of the unit halted at Höhr-Grenzhausen, where the 1st Company of Heavy Panzer Battalion (SS) 501 (with 13 Tiger I tanks) and parts of Jagdtiger Unit 512 were subordinated to it.

In March and April the unit was assigned to Reich Highway 8 to defend Siegen. Next it moved to the Schmallenberg area, where its last tanks were abandoned. The unit was disbanded.

Left: This tank was hit in its left front running gear near Freialdenhoven and was given up. The US photo caption "Shot down by an M-36 pursuit tank" (hit on the right side of the turret) is erroneous. Many immobilized and abandoned tanks were later hit by numerous shells, which the Allied crews then reported as kills.

Above: In a battle northeast of Bastogne, this tank was hit by eight antitank shells near Moinet; the crew was able to escape.

506

Below and lower left:
These two unpublished photos show King Tigers of the unit advancing in the direction of Bastogne.

Above:
Only a few photos exist of Tank Unit 509, which was established early in 1945. This one shows a tank in action in Hungary.

Below:
This photo shows a recovery tank towing one of the unit's vehicles in the German-Hungarian border area.

Above:
Units 502 and 507 received only limited numbers of the new Tiger II tanks. This photo shows the last tank used in the escape from a pocket at Halbe where it was put out of action. Meanwhile Unit 502 was renumbered 511.

509

Heavy Panzer Battalion 509

This unit was equipped with Tiger II tanks at Sennelager in mid-January 1945 and transferred to Hungary, south of Stuhlweissenburg. The attack in the direction of Budapest began partially without infantry support, and high losses resulted.

At the end of January the scattered unit had to defend itself against superior enemy attacks. Its strength — by this time reduced by five Tigers — amounted to only twelve Tiger II tanks then.

On January 31 the enclosed 3rd Panzer Division was relieved near Dunapentele. Early in February its sixteen tanks attacked Börgond to prevent a breakthrough of the Margarete position. This was successful. Bitter fighting in the Stuhlweissenburg area followed. The particularly capable repair shop was able to increase the number of active tanks again and again. During March the unit stood out in the heavy retreat fighting north of the Plattensee, and withdrew, fighting constantly, to the German border at Heiligenkreuz. Unfortunately, 14 Tiger II tanks had to be blown up because of fuel shortage.

Until early April what remained of the unit fought east of Lafnitz and was then transferred by rail to the St. Pölten area in the Mur Valley. Fighting in the Mailberg-Gross Harras area followed until early May. On May 7 the withdrawal to the Moldau began, and on the evening of May 8 the last five Tiger II tanks were blown up. The crews were taken prisoner by American troops on May 9.

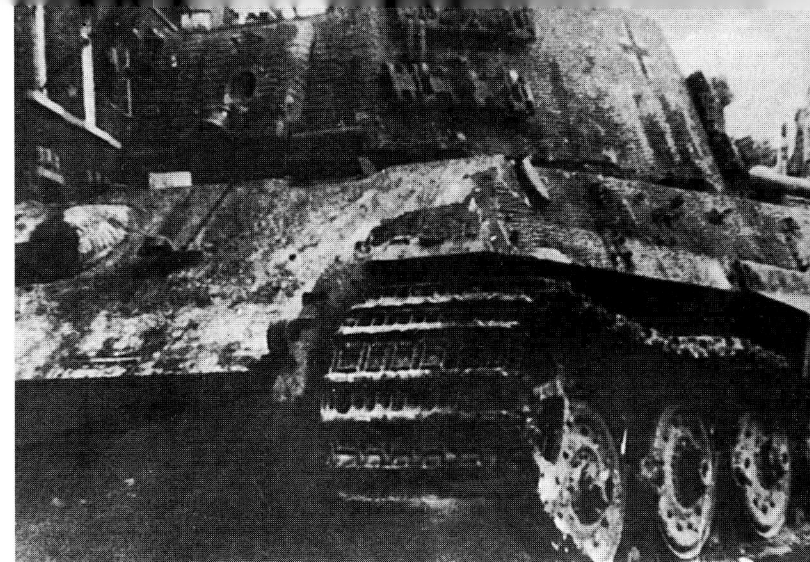

Above and upper right:
The 1st Company of SS Tank Unit 101 (later 501) was equipped with new King Tigers and saw service in eastern France. The two upper photos show Tank 101 with damage to its right running gear, and Tank 113.

Right:
Tank 104 was abandoned near Beauvais and later salvaged by the British. It is kept at Shrivenham today. (Not far from it, Tank 121 was disabled; it is now at the Tank Museum in Munster).

Upper left and right:
These two photos show a scene from a weekly newsreel: the marching column of SS Tank Unit 501 passes through the village of Tondorf on its way to the assembly area for the Ardennes offensive. The picture shows Staff Tiger 003; notice the "G" on the bow plate.

Left:
Later the 2nd Company followed; this picture shows Tank 222.

501 (SS)

Heavy SS Panzer Battalion 501

While still being set up, this unit was already subordinated to the 1st SS Panzer Division and took part in the Ardennes offensive since December 16, 1944 with the Peiper Battle Group.

Decimated again and again by breakdowns — usually of the technical kind — the battle group reached the La Gleize-Stoumont area on December 18-19 and was surrounded on December 20. Inadequately supplied from the beginning, the battle group had to leave its vehicles behind and fight its was through to its own troop on foot. All the unit's Tiger II tanks used there were lost. The unit's other tanks were used at first to secure the left flank in the Recht area, and later saw action at Vielsalm.

This 222 was photographed at several places on the advance route of the Peiper Battle Group: while passing a column of prisoners near Lanzerath (above) . . . with paratroopers of Fallschirmjäger Regiment 9 in Deidenberg (upper right) . . . and guarding a crossroads on Route 23 near St. Vith (right).

At the beginning of the new year, withdrawal fighting in the area before the Westwall ensued; in the process, the unit was often applied in pieces, down to single tanks (!). Then the unit was transferred to Hungary and, at the Gran bridgehead with the 1st SS Panzer Corps, was immediately involved in an attack on Fünfkirchen, where it had to abandon its last tanks in the tangle of retreat movement. Before that, it had already had to give up its 1st Company to Heavy Panzer Battalion 506.

There was daily fighting around the turning point of Stavelot. Tank 222 was hit on the left side of the hull while advancing toward the Ablève Bridge and had to be abandoned.

The lead tank of the 1st Company took several frontal hits on the Rue Haute-Rivage in Stavelot and, while moving backward, collided with house No. 9. The tank was stuck in the falling ruins.

Above:
Of the total of four King Tiger tanks left behind at La Gleize, two were in the Werimont Farm Group; here No. 211 is to be seen.

Upper left:
In the center of La Gleize, at the foot of the Chemin Vieille-Voie, another Tiger II remained after the end of the battle, which was used by soldiers of the U.S. 82nd Airborne Division for bazooka practice on January 18, 1945. To their amazement, the hollow charges did not penetrate the front armor.

Left:
A tank of the 3rd Company threw its right track on Route 33 near Intersection K 22 (1200 meters before La Gleize) and could not be made mobile. Several days later, American soldiers inspected the tank and set off a hand grenade in the turret.

The other tank, No. 213, rusted away until July 1945, when at the people's request the Americans towed it to the village square (upper right). From there it was moved by the Belgian Army to its showplace above the church in August of 1951. In 1975 it was given a non-authentic camouflage paint job (right). Only in 1981 was it given a more accurate one.

Tiger 332 was abandoned by its crew near Petit Coo and transported to Spa by rail by the 463rd Ordnance Evacuation Company (upper left). From there on it was taken to Antwerp and shipped to the U.S.A., where it was examined thoroughly at the Aberdeen Proving Ground (left). After the openings were closed, it was displayed (unfortunately with an inaccurate paint job).

Tiger 204 was stranded in a field without fuel. After being refueled, the tank was driven three kilometers in the direction of the Roanne-Coo railroad station by US soldiers. But it gave up the ghost on an upgrade and stayed there for weeks, an obstacle to traffic, before it was pushed down the slope.

North of La Gleize, Tiger 334 was used to secure Route 33 in the direction of Borgoumont. It slowed up the leading American tanks but was hit in its right running gear. At right is the wreck of a destroyed M 4 A 1 (76) tank. To clear the road, it was pushed into the ditch along the road once the tracks were detached.

Above:
Beyond Stavelot Tank 003, also seen in the Tondorf photos (page 34), was hit in the running gear. Before the crew abandoned it, they blew the muzzle brake so that the cannon would not move forward again after its last shot.

Right:
Tank 312 was abandoned near Goronne west of Vielsalm and fell into the hands of American troops.

Heavy SS Panzer Battalion 502

It was formed at Sennelager as of November 1944. Early in March of 1945 it was transported to the eastern front near Stettin and unloaded there on March 11. On March 18 it was taken by train to Müllrose, in the vicinity of Frankfurt an der Oder, and from there it attacked at the Küstrin bridgehead on March 21 and fought near Sachsendorf. The attack came to a stop, since the infantry could not follow.

On March 26 the unit advanced again as a (strengthened) battle group, but this attack also had to be broken off, because of enemy superiority and mines.

Early in April, retreat fighting took place against the Soviets, who attacked from the bridgehead. In mid-April there was fighting in the Lietzen-Marxdorf-Hasenfelde-Arensdorf area; there the unit was surrounded, but broke free through Wilmersdorf. At the end of April it fought on the north side of the Wolzinger See, and then broke out of the pocket of Halbe, gradually losing all its tanks in the process.

There are, unfortunately, very few pictures of what remained of SS Tank Units 502 (left) and 503 (opposite page), since these units were rushed into action in small groups.

Heavy SS Panzer Battalion 503

This unit, established as 1944 turned to 1945, was shipped to the eastern front on January 27, 1945, but never went into action as a unit. One part was sent to the Arnswalde area in Pomerania, the other to Landsberg-Küstrin.

The first group was surrounded at Arnswalde on February 4 and relieved by forces of the III. SS Tank Corps on February 12.

As plans were being made to apply the whole unit at Danzig under the command of the 2nd Army, the Soviet breakthrough toward Stettin took place and divided the unit. Only one half arrived at Danzig and was divided into four groups and subordinated to infantry divisions. The more and more decimated unit fought on until the end of March. A small part was shipped by sea, via Swinemünde, to the Berlin area (six tanks). Here the remainder of the unit, some of it fighting as infantry, was defeated.

Right:
The battalion commander (Sturmbannführer Herzig) awarding decorations.

Right:
One of the most successful commanders in the unit, Untersturmführer Karl Bromann, poses by the gun of his tank. The number of score rings speaks for itself.

Above: American officers occupy the works and examine turrets set on blocks.

Upper left:
Delivery of King Tigers was seriously delayed by a heavy air attack on the Henschel production plant in Kassel on October 7, 1944.

Left:
This photo shows finished turret and hull units on the Henschel firm's grounds.

The end: Made useless by barrel damage, this abandoned Tiger (with Porsche turret) waits to be scrapped on the grounds of Tank Unit 500 in Paderborn. To the right of the tank is a Panther V tank, and diagonally behind it the turret weight used on the VK 4501 (H) can be seen.